I0485235

Math Bridge

A Study Guide for
First Semester Calculus

Julean Albidone

This book and the information contained herein are for informative purposes only. The Information in this book is distributed on an as-is basis, without warranty. The author makes no legal claims, express or implied, and the material is not meant to substitute legal or financial counsel.

The author, publisher, and/or copyright holder assume no responsibility for the loss or damage caused or allegedly caused, directly or indirectly, by the use of information contained in this book. The author and publisher specifically disclaim any liability incurred from the use or application of the contents of this book.

Copyright © 2014 Julean Albidone

All rights reserved.

ISBN-10: 1499530609

ISBN-13: 978-1499530605

For my family.

Math Bridge is designed to teach the ordinary student how to succeed in Calculus 101. It strips away all the heavy theory and leaves you with the laymen's version, explained in plain English! This text is littered with visuals and examples; as through deep research and personal experience, Julean finds that this is the best way to fully understand a topic and ramp up quickly.

You can use this book as a study guide, a quick refresher or even a quick cheat sheet before a quiz! Don't get bogged down by heavy theory and proofs, get to understanding calculus quickly.

If you are a *Straight A* student and forced to take Calculus 101, then this is the book for you!

About the Author

As a former math teacher, tutor and test marker for Kumon, the world's largest after school math and reading program, **Julean Albidone** has taught hundreds of aspiring students. Although Julean never thought he was gifted in math, he was always at the top of his class and recipient of numerous calculus awards. He achieved this by breaking down the most complex topics into simple, small, bite sized chucks. Today he is a consultant with a background in mechanical engineering.

Table of Contents | 0.2

The Derivative Function | **1.1**

The derivative is used to determine how a function is changing over time. For example, if you have a position function and you wish to determine how quickly your position changes at a specific time, then you take the derivative; this will tell you the rate of change of position, or your velocity at that time. Further more, if you wish to determine how your velocity is changing, then you would take the derivative again to determine the acceleration. The derivative function is given below:

$$f'(x) = \lim_{h \to 0} \frac{f(x+h) - f(x)}{h}$$

That is, *f(x)* is the original function and *f'(x)* [spoken as: "**f** prime" or the "**derivative of f with respect to x**"] is the derivative of said function.

Graphically, the derivative of a function at a specific x value is represented as a tangent to the function at that co-ordinate. Below is the rate of change of $f(x) = x^3$ at various positions on the x axis:

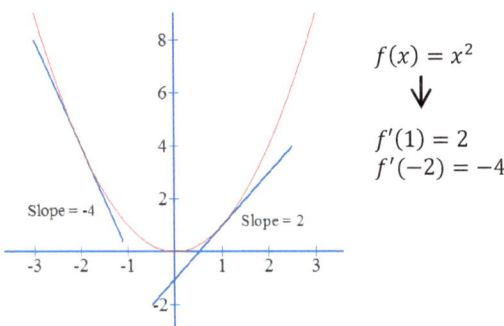

$f(x) = x^2$

$\downarrow$

$f'(1) = 2$
$f'(-2) = -4$

Example 1

Question: Find the derivative of $f(x) = x^2 + 2$,

Solution:

$$f'(x) = \lim_{h \to 0} \frac{f(x+h) - f(x)}{h}$$

1

$$f'(x) = \lim_{h \to 0} \frac{[(x+h)^2 + 2] - [x^2 + 2]}{h}$$

$$f'(x) = \lim_{h \to 0} \frac{[x^2 + 2xh + h^2 + 2] - [x^2 + 2]}{h}$$

$$f'(x) = \lim_{h \to 0} \frac{2xh + h^2}{h} = \lim_{h \to 0} 2x + h = 2x$$

* Note that the constant "+ 2" has no effect on the rate of change.

Example 2

Question: Find the derivative of $f(x) = x^2 + x$, then use this function to find the slope of the tangent line at $x = 1$ and $x = -3$.

Solution:

$$f'(x) = \lim_{h \to 0} \frac{[(x+h)^2 + (x+h)] - [x^2 + x]}{h}$$

$$f'(x) = \lim_{h \to 0} \frac{[x^2 + 2xh + h^2 + (x+h)] - [x^2 + x]}{h}$$

$$f'(x) = \lim_{h \to 0} \frac{2xh + h^2 + h}{h}$$

$$f'(x) = \lim_{h \to 0} 2x + h + 1 = 2x + 1$$

$$\therefore f'(1) = 2(1) + 1 = 3$$

$$\therefore f'(-3) = 2(-3) + 1 = -5$$

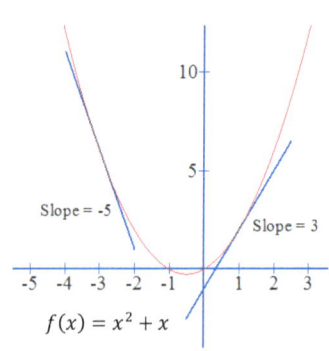

Slope = -5   Slope = 3

$f(x) = x^2 + x$

Techniques of Differentiation | 1.2

The last section covered the derivative on a fundamental level, giving us a very logical process of how any given function is changing with respect to time. But mathematicians have developed more efficient theorems that will allow us to arrive at the same "answer" with much less time and work devoted. This in turn allows us to derive more and more complex functions that would have otherwise been insurmountable following the techniques described in the previous chapter.

This section will thus outline various techniques for deriving various types of functions:

The Constant Function

The derivative of any constant "c" is zero, because the function remains at the same position in time indefinitely, the function is not changing, thus the derivative (or the functions rate of change) is zero:

$$\frac{d}{dx}[c] = 0$$

The Power Function

Any power function of any real number "r" can be derived by decreasing the exponent by one and then multiplying the power function by the original exponent:

$$\frac{d}{dx}[x^r] = rx^{r-1}$$

Constant times a Function

It is important to note that any constants times a function has no effect on the derivative. Thus we can factor out the constant from the function and then carry out the necessary steps for derivation:

$$\frac{d}{dx}[c \cdot f(x)] = c \cdot \frac{d}{dx}[f(x)] \qquad OR \qquad (cf)' = c(f)'$$

The Derivative of Sums and Differences

A function that is the combination of the sum or difference of two or more smaller functions can be dealt by first deriving its individual parts (given the functions are differentiable at x), the example is given for functions f and g:

3

$$(f + g)' = f' + g'$$
$$(f - g)' = f' - g'$$

Example 1

Question: Find the derivative of the various functions.

$$\frac{d}{dx}[5] = 0$$

$$\frac{d}{dx}[4x^7] = 4\frac{d}{dx}[x^7] = 4[(7)x^{7-1}] = 4[(7)x^6] = 28x^6$$

$$\frac{d}{dx}\left[\frac{1}{x^5}\right] = \frac{d}{dx}[x^{-5}] = [(-5)x^{(-5)-1}] = -5x^{-6} = -\frac{5}{x^6}$$

$$\frac{d}{dx}[x^2 + x^4] = 2x + 4x^3$$

$$\frac{d}{dx}\left[x^{2/3}\right] = \frac{2}{3}x^{(2/3)-1} = \frac{2}{3}x^{-1/3}$$

$$\frac{d}{dx}[\sqrt[4]{x}] = \frac{d}{dx}\left[x^{1/4}\right] = \frac{1}{4}x^{-3/4} = \frac{1}{4\sqrt[4]{x^3}}$$

Example 2

Question: Determine at what points the following function of $y = x^3 - 4x + 7$ has horizontal tangent lines.

Solution: We know from the previous chapter that a horizontal tangent line occurs when there is zero rate of change, or when the derivative of the function is zero. Thus, we must first take the derivative of the function:

$$y = x^3 - 4x + 7$$
$$y' = 3x^2 - 4$$

We must now find the points at which the derivative is zero, we do this by setting the value of y' to zero:

$$y' = 3x^2 - 4$$
$$0 = 3x^2 - 4$$
$$4 = 3x^2$$

4

$$\frac{4}{3} = x^2$$

$$x = \pm \sqrt{\frac{4}{3}}$$

$$x = \pm \frac{2}{\sqrt{3}}$$

Tips & Common Mistakes

Higher order derivatives, such as to the second (f'') or third (f''') order are carried out by deriving the derivative the desired amount of times. Fourth order derivatives and higher are written with roman numerals like so; fourth - $f^{(iv)}$, fifth - $f^{(v)}$, and so on.

The Product & Quotient Rules | 1.3

The Product Rule

If a function can be seen as the product of two individual functions, say f and g, then if the two functions are derivable for all real x, then so is their product. The Product rule is given below:

$$(f \cdot g)' = f \cdot g' + g \cdot f'$$

The Quotient Rule

If a function can be seen as the quotient of two individual functions, say f and g, then if the two functions are derivable for all real x, then so is their quotient:

$$\left(\frac{f}{g}\right)' = \frac{g \cdot f' - f \cdot g'}{g^2}$$

Example 1

Question: Derive the function $y = (3x^3 + 5)(x^2 - x)$ using the Product rule.

Solution:

$$y' = (3x^3 + 5)\frac{d}{dx}[(x^2 - x)] + (x^2 - x)\frac{d}{dx}[(3x^3 + 5)]$$
$$y' = (3x^3 + 5)(2x - 1) + (x^2 - x)(9x^2)]$$
$$y' = (6x^4 + 10x - 3x^3 - 5) + (9x^4 - 9x^3)$$
$$y' = (15x^4 - 12x^3 + 10x - 5)$$

Example 2

Question: Derive the function $y = \dfrac{x^2 + 3x + 5}{x - 3}$ using the Quotient rule.

Solution:

$$y' = \frac{(x - 3)(x^2 + 3x + 5)' - (x^2 + 3x + 5)(x - 3)'}{(x - 3)^2}$$
$$y' = \frac{(x - 3)(2x + 3) - (x^2 + 3x + 5)(1)}{x^2 - 6x + 9}$$

6

$$y' = \frac{(2x^2 - 3x - 9) - (x^2 + 3x + 5)}{x^2 - 6x + 9}$$

$$y' = \frac{x^2 - 6x - 4}{x^2 - 6x + 9}$$

Tips & Common Mistakes

- Do not make the mistake of deriving the functions separately first and then multiplying the two derivatives, the derivative of a product or quotient cannot be dealt in the same way that the sum or difference of two or functions can be.

- Sometimes it is easier to simplify a function first instead of applying the product or quotient rule straight away.

Trigonometric Functions | 1.4

This section will outline the derivatives of the six main trigonometric functions; sin*(x)*, cos*(x)*, tan*(x)*, cot*(x)* sec*(x)* and csc*(x)*:

$$(sinx)' = cosx \qquad (secx)' = secx \cdot tanx$$

$$(cosx)' = -sinx \quad (cotx)' = -csc^2x$$

$$(tanx)' = sec^2x \quad (cscx)' = -cscx \cdot cotx$$

Example 1

Question: Derive the function $y = (2x^2)cosx$

Solution:
$$y' = (2x^2)(cosx)' + (2x^2)'(cosx)$$
$$y' = (2x^2)(-sinx) + (4x)(cosx)$$
$$y' = 4xcosx - 2x^2sinx$$

Example 2

Question: Derive the function $y = cosx \cdot sinx$

Solution:
$$y' = (cosx)(sinx)' + (sinx)(cosx)'$$
$$y' = (cosx)(cosx) + (sinx)(-sinx)$$
$$y' = cos^2x - sin^2x$$

Example 3

Question: Derive the function $y = \frac{cosx}{1+sinx}$

Solution:
$$y' = \frac{cosx}{1+sinx}$$
$$y' = \frac{(1+sinx)(cosx)' - (cosx)(1+sinx)'}{(1+sinx)^2}$$
$$y' = \frac{(1+sinx)(-sinx) - (cosx)(cosx)}{(1+sinx)^2}$$
$$y' = \frac{(-sinx - sin^2x) - (cos^2x)}{(1+sinx)^2}$$

$$y' = \frac{-sinx - 1}{(1 + sinx)^2}$$
$$y' = -\frac{1}{1 + sinx}$$

The Chain Rule | 1.5

The chain rule applies when you are deriving a composite function such as $f \circ g$ [pronounced f at g] where both the functions f and g are differentiable at x. A composite function is when one function is evaluated at another, for example, $f(g(x))$, is the function of f evaluated at g. Below is the Chain rule:

$$\frac{d}{dx}[f(g(x))] = (f \circ g)'$$
$$= f(g(x))' \cdot g'$$

OR

$$\frac{dy}{dx} = \frac{dy}{du} \cdot \frac{du}{dx}$$
$$y = f(g(x)) \ and \ u = g(x)$$

Example 1

Question: Derive the function $y = sin\ (x^2)$

Solution: Assume $f = sin\ (x)$ and $g = x^2$, thus we arrive at:

$$y' = f'(x^2) \cdot (x^2)'$$
$$y' = cos(x^2) \cdot (2x)$$
$$y' = 2x\ cos(x^2)$$

OR

$$\frac{dy}{dx} = \frac{dy}{du} \cdot \frac{du}{dx}$$
$$\frac{dy}{dx} = \frac{d}{du}[sin\ u] \cdot \frac{d}{dx}[x^2]$$
$$\frac{dy}{dx} = \frac{d}{du}[sin\ u] \cdot \frac{d}{dx}[x^2]$$
$$\frac{dy}{dx} = cos\ u \cdot 2x = 2x\ cos\ x^2$$

Example 2

Question: Derive the function $y = (x^2 + csc\ x)^{2/3}$

Solution: Assume $f = (x)^{2/3}$ and $g = x^2 + csc\ x$, thus we arrive at:

$$y' = f'(x^2 + cscx) \cdot (x^2 + csc\ x)'$$
$$y' = \frac{2}{3}(x^2 + cscx)^{-1/3} \cdot (2x - csc\ x\ cot\ x)$$

$$y' = {}^2\!/_{3\sqrt[3]{x^2 + cscx}}(2x - csc\,x\,cot\,x)$$
$$y' = (4x - 2csc\,x\,cot\,x)\Big/{3\sqrt[3]{x^2 + cscx}}$$

Example 3

Question: Derive the function $y = {}^1\!/_{\sqrt{x^3 + cos\,x}}$

Solution: Rewrite the function as $y = (x^3 + cos\,x)^{-1/2}$, assume $f = (x)^{-1/2}$ and $g = x^3 + cos\,x$, thus we arrive at:

$$y' = f'(x^3 + cos\,x) \cdot (x^3 + cos\,x)'$$
$$y' = -\frac{1}{2}(x^3 + cos\,x)^{-3/2} \cdot (3x^2 - sin\,x)$$
$$y' = \frac{-1}{2}\frac{(3x^2 - sin\,x)}{\sqrt{(x^3 + cos\,x)^3}}$$
$$y' = \frac{sin\,x - 3x^2}{2\sqrt{(x^3 + cos\,x)^3}}$$

Implicit Differentiation │ 2.1

The previous chapter dealt with deriving functions of the form $y = f(x)$, but this greatly limits us to the number of functions we can differentiate. Implicit differentiation will allow us to differentiate functions for which it is difficult or impossible to express them in this form.

Example 1

Question: Derive the function $x^2y = 1$ by both implicit differentiation and conventional methods.

Solution 1 (Conventional):

$xy = 1$

$y = x^{-1}$

$y' = -x^{-2}$

$y' = -\dfrac{1}{x^2}$

Solution 2 (Implicit):

This is conducted by differentiating both sides with respect to x before solving for y in terms of x, we treat y as a differentiable function of x.

$xy = 1$

$\dfrac{d}{dx}xy = \dfrac{d}{dx}1$

$x\dfrac{d}{dx}y + y\dfrac{d}{dx}x = 0$

$x\dfrac{dy}{dx} + y = 0$

$\dfrac{dy}{dx} = -\dfrac{y}{x}$ *And we know that* $y = \dfrac{1}{x}$

$\therefore \dfrac{dy}{dx} = -\dfrac{1}{x^2}$

Example 2

Question: Derive the function $3y^3 + tan\,y = x^3$

Solution:

$$\frac{d}{dx}\left(3y(x)^3 + tan\,y(x)\right) = \frac{d}{dx}x^3$$

$$3\frac{d}{dx}[y(x)^3] + \frac{d}{dx}[tan\,y(x)] = \frac{d}{dx}x^3$$

Chain Rule: $\frac{d}{dx}[y(x)^3]$
$\frac{d}{dx}[y(x)^3] \to \frac{du^3}{du} \cdot \frac{du}{dx}$
$u = y(x)$
$= 3u^2 \cdot \frac{d}{dx}y(x)$
$= 3y^2 \cdot \frac{dy}{dx}$

Chain Rule: $\frac{d}{dx}[tan\,y(x)]$
$\frac{d}{dx}[tan\,y(x)] \to \frac{d\,tan\,u^3}{du} \cdot \frac{du}{dx}$
$u = y(x)$
$= sec^2\,u \cdot \frac{d}{dx}y(x)$
$= sec^2\,y \cdot \frac{dy}{dx}$

$$9y^2 \cdot \frac{dy}{dx} + sec^2\,y \cdot \frac{dy}{dx} = \frac{d}{dx}x^3$$

$$9y^2 \cdot \frac{dy}{dx} + sec^2\,y \cdot \frac{dy}{dx} = 3x^2$$

Now Rearrange:

$$\frac{dy}{dx} = \frac{3x^2}{9y^2 + sec^2\,y}$$

Note that the final derivate of $\frac{dy}{dx}$ is a function of both x and y.

Example 3

Question: Derive the function $x^3 + y^3 = xy$

Solution:

$$\frac{d}{dx}(x^3 + y^3) = \frac{d}{dx}(xy)$$

$$\frac{d}{dx}[x^3] + \frac{d}{dx}[y^3] = \frac{d}{dx}(xy)$$

$$3x^2 + 3y^2\frac{dy}{dx} = \frac{d}{dx}(xy)$$

> Use the Product Rule for $\frac{d}{dx}(xy)$
>
> $\frac{d}{dx}(xy) = x\frac{dy}{dx} + y(x)\frac{d}{dx}x$
>
> $\frac{d}{dx}(xy) = x\frac{dy}{dx} + y(x)$

$$3x^2 + 3y^2\frac{dy}{dx} = x\frac{dy}{dx} + y(x)$$

$$3x^2 + 3y^2\frac{dy}{dx} = x\frac{dy}{dx} + y(x)$$

$$3y^2\frac{dy}{dx} - x\frac{dy}{dx} = y(x) - 3x^2$$

$$-\frac{dy}{dx} = \frac{y(x) - 3x^2}{3y^2 - x}$$

$$\frac{dy}{dx} = \frac{3x^2 - y(x)}{x - 3y^2}$$

14

Logarithmic Functions | **2.2**

When differentiating a function, the natural logarithm is preffered over logarithms with other bases for simplicity reasons.

$$\frac{d}{dx}[\ln x] = \frac{1}{x}, x > 0$$

$$\frac{d}{dx}[\log_b x] = \frac{1}{x \ln b}, x > 0$$

Additionally, the chain rule can also apply to logarithmic functions:

$$\frac{d}{dx}[\ln u] = \frac{1}{u} \cdot \frac{du}{dx}$$

$$\frac{d}{dx}[\log_b u] = \frac{1}{u \ln b} \cdot \frac{du}{dx}$$

Example 1

Question: Derive the function $ln(x^3 + 4x^2)$

Solution: Let $u = x^3 + 4x^2$

$$y' = \frac{1}{u} \cdot \frac{du}{dx}$$
$$y' = \frac{1}{x^3 + 4x^2} \cdot \frac{d}{dx}[x^3 + 4x^2]$$
$$y' = \frac{1}{x^3 + 4x^2} \cdot [3x^2 + 8x]$$
$$y' = \frac{3x^2 + 8x}{x^3 + 4x^2} = \frac{3x + 8}{x^2 + 4x}$$

Example 2

Question: Derive the function $\log_5[10x + x^2]$

Solution: Let $u = 10x + x^2$

$$y' = \frac{1}{u \ln b} \cdot \frac{du}{dx}$$

$$y' = \frac{1}{[10x + x^2]\ln 5} \cdot \frac{d}{dx}[10x + x^2]$$

$$y' = \frac{10 + 2x}{[10x + x^2]\ln 5}$$

Example 3

Question: Derive the function:

$$y = \ln\frac{x^2 \cos x}{\sqrt{5 + x}}$$

Solution: Rewrite the function in the form $\ln x^2 + \ln(\cos x) - \ln\sqrt{5 + x}$

$$y = 2\ln x + \ln(\cos x) - \frac{1}{2}\ln(5 + x)$$

$$y' = 2\frac{1}{x} + \frac{1}{\cos x}[-\sin x] - \frac{1}{2}\frac{1}{5+x} \cdot [1]$$

$$y' = \frac{2}{x} - \tan x - \frac{1}{10+2x}$$

Example 4

Question: Derive the function:

$$y = \frac{x^3\sqrt{3x + 5x^2}}{(1 + x^2)^3}$$

Solution: First multiply the function by the natural logarithm and then use its properties to break the function down into one that is more manageable.

$$\ln y = \ln\frac{x^3\sqrt{3x + 5x^2}}{(1 + x^2)^3}$$

$$\frac{dy}{dx}[\ln y] = \frac{dy}{dx}[3\ln x + \frac{1}{2}\ln(3x + 5x^2) - 3\ln(1 + x^2)]$$

$$\frac{1}{y}\frac{dy}{dx} = \frac{3}{x} + \frac{3 + 10x}{2(3x + 5x^2)} - \frac{6x}{1 + x^2}$$

$$\frac{dy}{dx} = y[\frac{3}{x} + \frac{3 + 10x}{2(3x + 5x^2)} - \frac{6x}{1 + x^2}]$$

$$\frac{dy}{dx} = \frac{x^3\sqrt{3x + 5x^2}}{(1 + x^2)^3}[\frac{3}{x} + \frac{3 + 10x}{2(3x + 5x^2)} - \frac{6x}{1 + x^2}]$$

Tips & Common Mistakes

- Keep in mind that $\frac{dy}{dx}\ln|x| = \frac{1}{x}, x \neq 0$, note that the derivative of $\ln|x|$ and $\ln x$, are the same, just that the absolute natural logarithm is restricted to $x \neq 0$.

Exponential & Inverse Trigonometric | 2.3

One to One Functions

The derivative of a one to one function is key in determining the derivative of its inverse function, (f^{-1}). A function has an inverse if and only if it is a one to one function. In general, if f is a one to one function and differentiable, then the derivative of its inverse function can be defined as:

$$(f^{-1})'(x) = \frac{1}{f'(f^{-1}(x))}$$

A Function is one to one on $(-\infty, +\infty)$ if:

$$f'(x) > 0 \text{ or } f'(x) < 0$$

OR

$f(x)$ *passes the horizontal line test (two or more lines cannot pass through any horizontal line, similar to vertical line test).*

Example 1

Question: a) Show that f is a one to one function, $f(x) = x^5 + 5x + 1$ b) Find the derivative of f^{-1}

Solution a):
$f'(x) = 5x^4 + 5, f' > 0$
Therefore the function is one to one.

Solution b): let $y = (f^{-1})(x)$

$$(f^{-1})'(x) = \frac{1}{f'(f^{-1}(x))}$$

$$y'(x) = \frac{1}{f'(y)} = \frac{1}{\frac{d}{dy}[y^5 + 5y + 1]}$$

$$\therefore y'(x) = \frac{1}{5y^4 + 5}$$

Exponential Functions

Given an exponential function of the form $b^x (b > 0, b \neq 1)$, then the derivative of the exponential function is as follows:

$$\frac{d}{dx}[b^x] = b^x \ln b$$

$$\frac{d}{dx}[b^u] = b^u \ln b \cdot \frac{du}{dx}$$

Additionally if **b = e** such that **ln e = 1**, then we have as follows:

$$\frac{d}{dx}[e^x] = e^x$$

$$\frac{d}{dx}[e^u] = e^u \frac{du}{dx}$$

Example 2

Question: Find

$$a)\ \frac{d}{dx}[5^x], \qquad b)\frac{d}{dx}[e^{\cos x}]$$

Solution a):

$$\frac{d}{dx}[5^x] = 5^x \ln 5$$

Solution b):

$$\frac{d}{dx}[e^{\cos x}] = e^{\cos x} \cdot \frac{d}{dx}[\cos x]$$

$$\frac{d}{dx}[e^{\cos x}] = -(\sin x)e^{\cos x}$$

Inverse Trigonometric Functions

The following is a complete list of the six basic inverse trigonometric functions, $sin^{-1} x, cos^{-1} x, tan^{-1} x, csc^{-1} x, sec^{-1} x$ **and** $cot^{-1} x$.

$$\frac{d}{dx}[sin^{-1} u] = \frac{1}{\sqrt{1-u^2}} \cdot \frac{du}{dx}$$

$$\frac{d}{dx}[cos^{-1} u] = -\frac{1}{\sqrt{1-u^2}} \cdot \frac{du}{dx}$$

$$\frac{d}{dx}[tan^{-1} u] = \frac{1}{1+u^2} \cdot \frac{du}{dx}$$

$$\frac{d}{dx}[csc^{-1} u] = -\frac{1}{|u|\sqrt{u^2-1}} \cdot \frac{du}{dx}$$

$$\frac{d}{dx}[sec^{-1} u] = \frac{1}{|u|\sqrt{u^2-1}} \cdot \frac{du}{dx}$$

$$\frac{d}{dx}[cot^{-1} u] = -\frac{1}{1+u^2} \cdot \frac{du}{dx}$$

Example 3

Question: Find

$$a)\ \frac{d}{dx}[cos^{-1}(x^3)] \text{ and } b)\frac{d}{dx}[csc^{-1}(5x^2)]$$

19

Solution *a*):

$$\frac{d}{dx}[cos^{-1}(x^3)] = -\frac{1}{\sqrt{1-x^6}} \cdot \frac{d}{dx}x^3$$

$$= -\frac{3x^2}{\sqrt{1-x^6}}$$

Solution *b*):

$$\frac{d}{dx}[csc^{-1}(5x^2)] = -\frac{1}{|5x^2|\sqrt{25x^4-1}} \cdot \frac{d}{dx}5x^2$$

$$= -\frac{10x}{|5x^2|\sqrt{25x^4-1}}$$

Tips & Common Mistakes

- A one to one function is a function that can be solved for each variable. For example; $y = 5x$ is one to one because it can be written as $x = \frac{y}{5}$, $y = x^2$ is not because x is not expressed as a single function of y ($x = \pm\sqrt{y}$).
- A functions of the form $f(x) = a^b$ in which both a and b are non-constant functions (for example $y = (x^3 + 5)^{cos\,x}$), is not an exponential function, functions of this form can be differentiated using logarithmic differentiation. See the previous chapter.

Related Rates | 2.4

There is no set formula or theorem that you can use to solve related rates problems, instead we can follow a set strategy for tackling these types of problems. This strategy will help us in breaking down the problem, coming up with a general equation and then deriving the equation to solve for the unknown rate of change. A general strategy is outlined below:

A General Strategy for Solving Related Rates Problems:

Step 1. Draw a diagram that illustrates the scenario.

Step 2. Assign variables to all quantities that change with time.

Step 3. Identify the rates of change that are known and that are to be determined. Each rate is a derivative.

Step 4. Find an equation that relates the known rates of change to the rate of change to be determined.

Step 5. Differentiate both sides of the equation to find the unknown rate of change.

Step 6. If necessary, substitute all known values and then solve (rearrange) for the unknown rate of change.

Example 1

Question: Plane A is 50km west of plane B. Plane A is traveling south at 180km/h and ship B is traveling north at 250km/h. How fast is the distance between the ships changing four hours from the initial starting time.

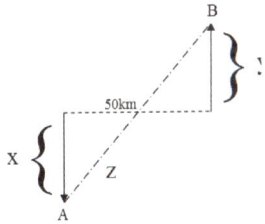

Solution:

x = the distance traveled by plane A in 4 hours,

$$\frac{dx}{dt} = 180$$

y = the distance traveled by plane B in 4 hours,

$$\frac{dy}{dt} = 250$$

z = the distance between the two planes

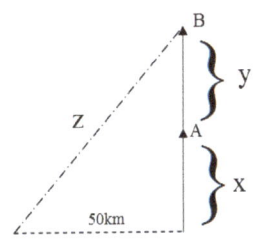

$$\frac{dz}{dt} = ?$$

We can redraw our diagram as a right triangle and then solve the equation using Pythagorean's theorem.

$$50^2 + (x + y)^2 = z^2$$

$$\frac{d}{dt}[50^2 + (x + y)^2] = \frac{dz}{dt}[z^2]$$

$$2(x + y)(\frac{dx}{dt} + \frac{dy}{dt}) = 2z\frac{dz}{dt}$$

$$\frac{(x + y)(\frac{dx}{dt} + \frac{dy}{dt})}{z} = \frac{dz}{dt} \quad \textbf{\textit{now sub in variables}}$$

$$\frac{dz}{dt} = \frac{(720 + 1000)(180 + 250)}{\sqrt{50^2 + (720 + 1000)^2}}$$

$$\frac{dz}{dt} = 429.82km/h$$

Example 2

Question: Gravel is being dumped at a rate of 2m³/min. The pile of gravel is such that it forms a cone whose diameter is always equal to its height. How fast is the height of the pile increasing if the pile is 5m high?

Solution:

$$h = 2r = d \qquad V = \frac{1}{3}\pi r^2 \cdot h$$

$$\frac{dV}{dt} = 2 \qquad V = \frac{1}{3}\pi(\frac{h}{2})^2 \cdot h$$

$$\frac{dh}{dt} = ? \qquad V = \frac{\pi}{12}h^3$$

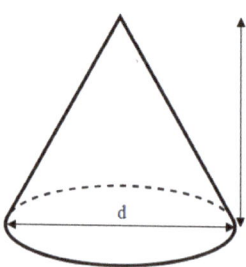

$$V = \frac{\pi}{12}h^3$$

$$\frac{d}{dt}V = \frac{d}{dt}[\frac{\pi}{12}h^3]$$

$$\frac{dV}{dt} = \frac{\pi}{4}h^2\frac{dh}{dt}$$

$$2 = \frac{\pi}{4}(5)^2\frac{dh}{dt}$$

$$\frac{dh}{dt} = 0.102m/min$$

Local Linear Approximation | 2.5

Recall that when you zoom in on any portion of a differentiable function, that the function takes the appearance of a straight line. This forms the basis of local linear approximation; we can approximate any x value of a function near a given x_0. Note, you might be asking yourself, why approximate a value of a function when you can simply find the exact x value by plugging it directly into the function itself, why come up with an entirely new approximation formula? The main application of local linear approximation is in modeling problems where it is useful to replace complicated functions by simpler ones.

$$f(x) \approx f(x_0) + f'(x_0)(x - x_0)$$

Example 1

Question: Find the slope using local linear approximation of $f(x) = x^2$ at the point $x_0 = 3$

Solution:

$$f(x) \approx f(x_0) + f'(x_0)(x - x_0)$$
$$f(x) \approx (3)^2 + [2(3)](x - 3)$$
$$f(x) \approx 9 + 6x - 18$$
$$f(x) \approx 6x - 9$$

Example 2

Question: *a)* Find the local linear approximation of $f(x) = \cos x$ at the point $x_0 = \frac{\pi}{2}$.

b) Use this function to approximate $f\left(\frac{2\pi}{5}\right)$, and then compare your result with the one produced directly by the function

Solution *a*):

$$f(x) \approx f(x_0) + f'(x_0)(x - x_0)$$
$$f(x) \approx \cos(\tfrac{\pi}{2}) - \sin(\tfrac{\pi}{2})\,(x - \tfrac{\pi}{2})$$
$$f(x) \approx (0) - (1)(x - \tfrac{\pi}{2})$$
$$f(x) \approx \frac{\pi}{2} - x$$

23

Solution b):

$$f\left(\frac{2\pi}{5}\right) \approx \frac{\pi}{2} - \frac{2\pi}{5} \qquad f\left(\frac{2\pi}{5}\right) = \cos\left(\frac{2\pi}{5}\right)$$

$$f\left(\frac{2\pi}{5}\right) \approx \frac{\pi}{10} \qquad\qquad f\left(\frac{2\pi}{5}\right) = 0.3090169943$$

      This makes sense that the approximated value is slightly higher than the one directly computed, see graph. Take note that the further away you are from x_0, the greater the error in the approximating function.

L'Hôpital's Rule, Indeterminate | 2.6

L'Hôpital's rule is a general method for using derivatives to find limits. There are many indeterminate forms of limits that we can analyze, this section will cover them all.

Indeterminate Forms of the Type 0/0 and ∞ / ∞

Suppose a limit of the form $\lim\limits_{x \to a} \frac{f(x)}{g(x)}$ is such that the limits of

$\lim\limits_{x \to a} f(x)$ and $\lim\limits_{x \to a} g(x)$ are both 0 or are both ∞, then the $\lim\limits_{x \to a} \frac{f(x)}{g(x)}$ then the

limit can be simplified to:

$$\lim_{x \to a} f(x) = 0 \text{ and } \lim_{x \to a} g(x) = 0$$

OR

$$\lim_{x \to a} f(x) = \infty \text{ and } \lim_{x \to a} g(x) = \infty$$

THEN

$$\lim_{x \to a} \frac{f(x)}{g(x)} = \lim_{x \to a} \frac{f'(x)}{g'(x)}$$

This statement holds true in the case of $x \to a^-$, $x \to a^+$, $x \to +\infty$ or as $x \to -\infty$. Additionally this statement only holds true if the limit of $\frac{f'(x)}{g'(x)}$ is finite, $+\infty$ or $-\infty$.

Example 1

Question: Find

$$a) \lim_{x \to 2} \frac{x^2 - 4}{x - 2}, \ b) \lim_{x \to 0^+} \frac{\ln x}{\csc x} \ and \ c) \lim_{x \to \frac{\pi}{2}} \frac{1 - \sin x}{\cos x}$$

Solution *a*):

$$\lim_{x \to 2} x^2 - 4 = 0$$
$$\lim_{x \to 2} x - 2 = 0$$

$$\therefore \text{ of the form } \frac{0}{0}$$

$$\lim_{x \to 2} \frac{x^2 - 4}{x - 2} = \lim_{x \to 2} \frac{d}{dx} \left[\frac{x^2 - 4}{x - 2} \right]$$

$$\lim_{x \to 2} \frac{x^2 - 4}{x - 2} = \lim_{x \to 2} \frac{2x}{1}$$

$$\lim_{x \to 2} \frac{x^2 - 4}{x - 2} = 4$$

Solution *b*):

$$\lim_{x \to 0^+} \ln x = \infty$$

$$\lim_{x \to 0^+} \csc x = \infty$$

$$\therefore \text{ of the form } \infty/\infty$$

$$\lim_{x \to 0^+} \frac{\ln x}{\csc x} = \lim_{x \to 0^+} \frac{d}{dx} \left[\frac{\ln x}{\csc x} \right]$$

$$\lim_{x \to 0^+} \frac{\ln x}{\csc x} = \lim_{x \to 0^+} \frac{1/x}{-\csc x \cot x}$$

But this is still of the form ∞/∞, if we apply L'Hôpital's rule again we will simply get powers of $1/x$ and expressions of $\csc x$ and $\cot x$ in the denominator, thus giving us more indeterminate forms. We must rearange and solve.

$$\lim_{x \to 0^+} \frac{1/x}{-\csc x \cot x} = \lim_{x \to 0^+} \frac{-\sin x}{x} \tan x$$

$$\lim_{x \to 0^+} \frac{-\sin x}{x} \tan x = -(1)(0) = 0$$

Solution *c*):

$$\lim_{x \to \frac{\pi}{2}} 1 - \sin x = 0$$

$$\lim_{x \to \frac{\pi}{2}} \cos x = 0$$

$$\therefore \text{ of the form } 0/0$$

$$\lim_{x \to \frac{\pi}{2}} \frac{1 - \sin x}{\cos x} = \lim_{x \to \frac{\pi}{2}} \frac{d}{dx} \left[\frac{1 - \sin x}{\cos x} \right]$$

$$\lim_{x \to \frac{\pi}{2}} \frac{1 - \sin x}{\cos x} = \lim_{x \to 2} \frac{-\cos x}{-\sin x}$$

$$\lim_{x \to \frac{\pi}{2}} \frac{1 - \sin x}{\cos x} = \frac{0}{-1} = 0$$

Indeterminate Forms of the Type $0 \cdot \infty$ and $\infty \cdot \infty$

Indeterminate forms of the type $0 \cdot \infty$ and $\infty \cdot \infty$ are also common, the following are examples of such:

$$\lim_{x \to \pi}(x - \pi)\cot x \qquad \text{and} \qquad \lim_{x \to 0^+}\left(\frac{1}{x} - \frac{1}{\sin x}\right),$$

respectively. These problems can often be solved by rearranging the limits into the indeterminate forms of $0/0$ and ∞/∞. For example, the limits above can both be rearranged into the form ∞/∞:

$$\lim_{x \to \pi}(x - \pi)\cot x \;\to\; \lim_{x \to \pi}\frac{\cot x}{\left(\dfrac{1}{x - \pi}\right)}$$

$$\lim_{x \to 0^+}\left(\frac{1}{x} - \frac{1}{\sin x}\right) \;\to\; \lim_{x \to \pi}\frac{\sin x - x}{x\sin x}$$

from here these limits can be solved by L'Hôpital's rule.

Indeterminate Forms of the Type 00, ∞^0, 1^∞

Indeterminate forms of this type can often be solved by first introducing a dependent variable to force the form $y = f(x)^{g(x)}$, then you can apply the natural logarithmic function, from here you solve for $\ln y$, as shown below:

$$y = f(x)^{g(x)}$$
$$\ln y = \ln f(x)^{g(x)}$$
$$\ln y = g(x)\ln f(x)$$
$$\lim_{x \to a}\ln y = \lim_{x \to a} g(x)\ln f(x)$$
$$\ln y = \lim_{x \to a} g(x)\ln f(x)$$

Example 2

Question: Find the limit of:

$$\lim_{x \to 0}(e^x + x)^{1/x}$$

Solution:

$$y = (e^x + x)^{1/x}$$
$$\ln y = \ln(e^x + x)^{1/x}$$

27

$$ln\,y = \frac{ln(e^x + x)}{x}$$

$$\lim_{x \to 0} ln\,y = \lim_{x \to 0} \frac{ln(e^x + x)}{x}$$

$$ln\,y = \lim_{x \to 0} \frac{e^x + 1}{e^x + x}$$

$$ln\,y = 2$$

$$e^{ln\,y} = e^2$$

$$y = e^2$$

$$\therefore \lim_{x \to 0}(e^x + x)^{1/x} = e^2$$

Tips & Common Mistakes

- Indeterminate forms of the type ∞ - ∞ lead to the following expressions:

$$(+\infty) - (+\infty), \qquad (-\infty) - (-\infty),$$
$$(+\infty) + (-\infty), \qquad (-\infty) + (+\infty).$$

They are indeterminate because the two terms exert conflicting positive and negative directions on the outcome of the limit. Limit problems that are not indeterminate are of the following:

$$(+\infty) - (-\infty), \qquad (+\infty) + (+\infty),$$
$$(-\infty) + (-\infty), \qquad (-\infty) - (+\infty).$$

Increase & Decrease, 3.1
Concavity

Increasing and Decreasing Functions

One sure-fire way to know if a function is decreasing is to differentiate the function and evaluate it at a certain x value, if the value is positive, it is increasing, if it is negative, the function is decreasing. You can also determine the intervals at which the function is increasing and decreasing by solving the differentiated function for x.

$$f'(x) > 0 \rightarrow Increasing$$
$$f'(x) < 0 \rightarrow Decreasing$$
$$f'(x) = 0 \rightarrow Constant$$

Solving f' for x
= Intervals

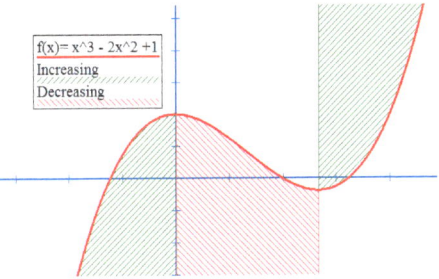

f(x)= x^3 - 2x^2 +1
Increasing
Decreasing

Concavity

Concavity reveals the direction of the curve. For example in the graph below, the center portion of the graph is decreasing, but the left half has a downward curve and the right half has an upwards curve. Another way to look at it is; for concave up, the slope is increasing, and for concave down, the slope is decreasing. The change in concavity is denoted by an inflection point. Concavity is found by taking the second derivative of the function.

$$f''(x) > 0 \rightarrow Concave\ up$$
$$f''(x) < 0 \rightarrow Concave\ down$$

Solving f'' for x
= Inflection Point

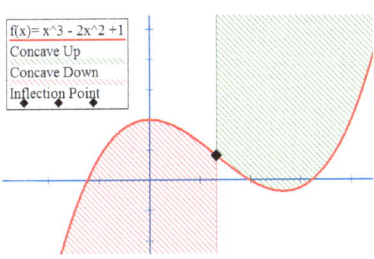

f(x)= x^3 - 2x^2 +1
Concave Up
Concave Down
Inflection Point

Example 1

Question: Find the intervals on which the function $f(x) = x^3 - 2x^2 + 1$ is increasing and decreasing. Also find the concavity and inflection point(s) of this function.

Solution:

$f(x) = x^3 - 2x^2 + 1$
$f'(x) = 3x^2 - 4x$
$f'(x) = x(3x - 4)$
Now solving for x
$x = 0 \text{ and } \dfrac{4}{3}$

Interval	$f'(x)$	Conclusion
$x < 0$	+	Increasing
$0 < x < 4/3$	−	Decreasing
$x > 4/3$	+	Increasing

$f'(x) = 3x^2 - 4x$
$f''(x) = 6x - 4$
Now solving for x
$x = \dfrac{2}{3}$
$\therefore$ *the inflection point is at* $\dfrac{2}{3}$

Interval	$f''(x)$	Conclusion
$x < 2/3$	−	Concave Down
$x > 2/3$	+	Concave Up

Tips & Common Mistakes

- To figure out if the derivative (or even the second derivative) is positive or negative, take any value along that integral and plug it back into the derivative.

Let's take the example above; we have arrived at the second derivative and found our inflection point. Now let's discover the concavity below.

For the interval $x < 2/3$, let's plug a value along that interval, let's say $x = 1/3$, into our second derivative of $f''(x) = 6x - 4$:

$$f''\left(\frac{1}{3}\right) = 6\left(\frac{1}{3}\right) - 4$$
$$f''\left(\frac{1}{3}\right) = 2 - 4$$
$$f''\left(\tfrac{1}{3}\right) = -2$$

This is less than zero and therefore it is Concave Down, as is seen above.

Relative Extrema | 3.2

Relative extrema are very common in functions, they occur when a function changes from increasing to decreasing, or vice versa. Relative maxima and minima are generally obtained two different ways:

$f'(a^-) > 0$ and $f'(a^+) < 0$,
then Maxima

$f'(a^-) < 0$ and $f'(a^+) > 0$,
then Minima

OR

$f'(x_0) = 0$ and $f''(x_0) > 0$,
then Minima

$f'(x_0) = 0$ and $f''(x_0) < 0$,
then Maxima

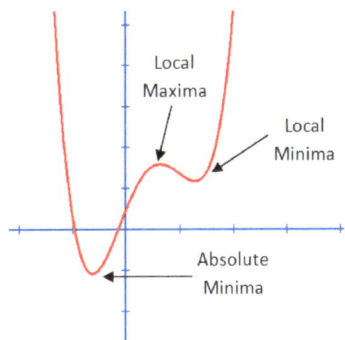

Local Maxima

Local Minima

Absolute Minima

Example 1

Question: Find the local maxima and/or minima of the function $f(x) = 4x^3 - 9x$

Solution:

$f(x) = 4x^3 - 9x$
$f'(x) = 12x^2 - 9$
Now solve for x,
$x = \pm \dfrac{3}{2\sqrt{3}}$

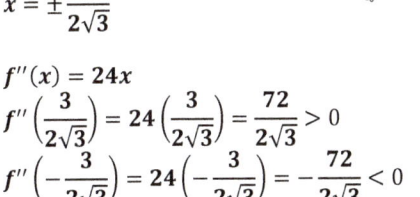

$f''(x) = 24x$
$f''\left(\dfrac{3}{2\sqrt{3}}\right) = 24\left(\dfrac{3}{2\sqrt{3}}\right) = \dfrac{72}{2\sqrt{3}} > 0$
$f''\left(-\dfrac{3}{2\sqrt{3}}\right) = 24\left(-\dfrac{3}{2\sqrt{3}}\right) = -\dfrac{72}{2\sqrt{3}} < 0$
$\therefore$ *Minima at* $x = \dfrac{3}{2\sqrt{3}}$,   *Maxima at* $x = -\dfrac{3}{2\sqrt{3}}$

Rational Functions and Cusps | 3.3

There are several characteristics of graphs that must be considered when graphing a function:

- symmetry
- x & y intercepts
- relative extrema
- intervals of increase and decrease
- asymptotes (rational functions)

- periodicity (trig functions)
- concavity
- inflection points
- as $x \to +\infty$ and as $x \to -\infty$

Rational Functions

Rational functions are functions of the form $f(x) = P(x)/Q(x)$ where $P(x)$ and $Q(X)$ are polynomials. Graphs of rational functions are often far more complex than polynomials because of the possibility of asymptotes and discontinuities.

Vertical asymptotes occur at values of x for which $Q(x) = 0$. For example, the function $f(x) = \dfrac{2x - 6}{4 - x}$ has a vertical asymptote at $x = 4$ because $4 - x = 0$ at this value. You can see this behaviour in the graph to the right.

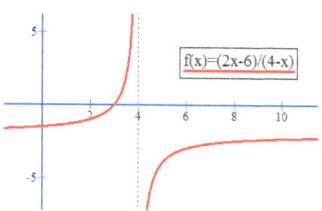

Horizontal asymptotes are determined by evaluating the end limits of the function at $x \to +\infty$ and at $x \to -\infty$. For example, if we evaluate the end behaviour of the function $f(x) = \dfrac{x^2}{1 - x^3}$:

$$\lim_{x \to +\infty} \frac{x^2}{1 - x^3} = \lim_{x \to +\infty} \frac{2x}{-3x^2} = \lim_{x \to +\infty} \frac{2}{-6x} = 0$$

$$\lim_{x \to -\infty} \frac{x^2}{1 - x^3} = \lim_{x \to -\infty} \frac{2x}{-3x^2} = \lim_{x \to -\infty} \frac{2}{-6x} = 0$$

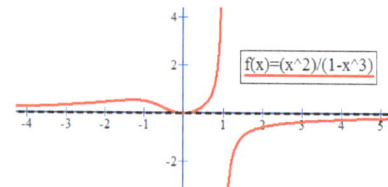

we can see that both the left and right limits are zero. This can be seen in the graph above.

Oblique and Curvilinear Asymptotes

Oblique and curvilinear asymptotes occur when the numerator of the function is of a higher degree than the denominator. It is often easier to analyze these types of asymptotes by rewriting these functions by dividing the numerator and denominator. For example; $f(x) = {(x^2 + 1)}/{x}$ can be rewritten in the form $f(x) = x + {1}/{x}$, as $x \to \pm\infty$ the $\frac{1}{x}$ term is forced to zero, and thus the function approaches $y = x$. See the graph on the right.

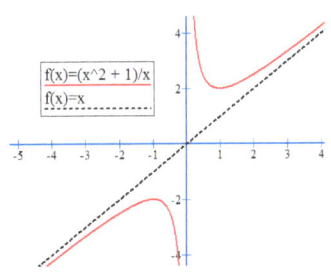

The previous example with the asymptote of $y = x$ is called an oblique asymptote. This next rational function is an example of a curvilinear asymptote:

$$f(x) = \frac{x^3 - 4x - 8}{x + 2}$$

$$f(x) = \frac{x^3 - 4x}{x + 2} - \frac{8}{x + 2}$$

$$f(x) = \frac{x(x^2 - 4)}{x + 2} - \frac{8}{x + 2}$$

$$f(x) = \frac{x(x + 2)(x - 2)}{x + 2} - \frac{8}{x + 2}$$

$$f(x) = x^2 - 2x - \frac{8}{x + 2}$$

Thus as $x \to +\infty$ and $-\infty$ the $\frac{8}{x+2}$ term approaches zero and the $x^2 - 2x$ becomes dominant, therefore the function approaches the curvilinear asymptote of $x^2 - 2x$. The graph confirms our answer. Also note the vertical asymptote at $x = -2$.

Oblique and curvilinear asymptotes:

When $f(x) = P(x)/Q(x)$ can be written as:

$$f(x) = q(x) + {r(x)}/{Q(x)} \text{ such that } {r(x)}/{Q(x)} \to 0 \text{ as } x \to \pm\infty$$

Vertical Tangents and Cusps

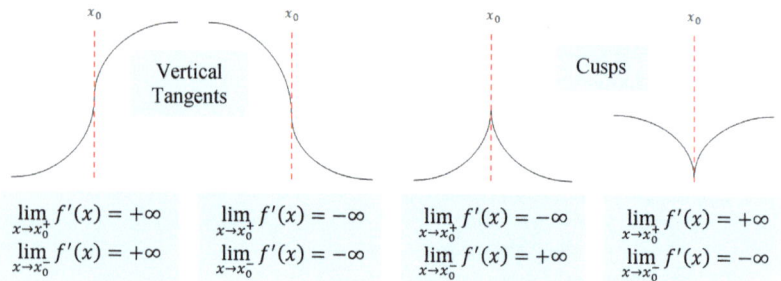

$$\lim_{x \to x_0^+} f'(x) = +\infty$$
$$\lim_{x \to x_0^-} f'(x) = +\infty$$

$$\lim_{x \to x_0^+} f'(x) = -\infty$$
$$\lim_{x \to x_0^-} f'(x) = -\infty$$

$$\lim_{x \to x_0^+} f'(x) = -\infty$$
$$\lim_{x \to x_0^-} f'(x) = +\infty$$

$$\lim_{x \to x_0^+} f'(x) = +\infty$$
$$\lim_{x \to x_0^-} f'(x) = -\infty$$

General Procedure for Graphing Rational Functions

Graphing a Rational Function, $f(x) = \dfrac{P(x)}{Q(x)}$

Step 1 – Symmetry	Determine if there is symmetry about the x and/or y axis.
Step 2 – Intercepts	Find the x and y intercepts.
Step 3 – Oblique and Curvilinear Asymptotes	See if the function can be written as $f(x) = q(x) + \dfrac{r(x)}{Q(x)}$ such that $\dfrac{r(x)}{Q(x)} \to 0$ as $x \to \pm\infty$.
Step 3.1 – Vertical Asymptotes	If there is no oblique or curvilinear asymptotes, then find the values of x such that $Q(x) = 0$.
Step 4 – Sign of $f(x)$	Calculate a range of x values to determine the intervals at which the function is positive and negative, this will tell you at what intervals the function is above or below the x axis.
Step 5 – Horizontal Asymptotes	Compute the limits of $f(x)$ as $x \to +\infty$ and at $x \to -\infty$. If either of the limits have a finite value, than this is the horizontal asymptote.
Step 6 – Derivatives	Compute both the first and second derivative to find the intervals of increase and decrease and to find the concavity and inflection points. Also use this to find the relative extrema. See if there is any value of x such that $f'(x) = 0$.
Step 7 – Vertical Tangents/Cusps	By analyzing the first derivative, see if there is any values of x that will force the function to $+\infty$ or $-\infty$. Then determine if this is a vertical tangent or a cusp.
Step 8 – Conclusion	Couple all the previous steps to form a complete overview of the graph. Sketch a graph that satisfies all of these conclusions.

Example 1

Question: Graph $f(x) = \dfrac{(x-2)^3}{x^2}$

Solution:

Symmetry: there is no symmetry about this function.
Intercepts: there are no y intercepts, but there are x intercepts:

$$0 = \frac{(x-2)^3}{x^2}$$
$$0 = (x-2)^3$$
$$x = 2$$

Oblique / Curvilinear Asymptotes: the function can also be written as

$$f(x) = x - \frac{8}{x^2} + \frac{12}{x} - 6$$

therefore as $x \to \pm\infty$ the function approaches $f(x) = x - 6$.

Vertical Asymptotes: there is one at $x = 0$:

$$Q(x) = x^2$$
$$0 = x^2$$
$$x = 0$$

Sign of $f(x)$: by taking a range of values, we find that the function is below the x-axis up until $x = 2$ where it then remains above the x-axis.
Horizontal Asymptotes: there are no horizontal asymptotes as the function goes to $+\infty$ and $-\infty$.

Derivatives:

Interval	$f'(x)$	Conclusion	Interval	$f''(x)$	Conclusion
$x < -4$	+	Increasing	$x < 0$	−	Conc. Down
$-4 < x < 0$	−	Decreasing	$0 < x < 2$	−	Conc. Down
$0 < x < 2$	+	Increasing	$x > 2$	+	Conc. Up
$x > 2$	+	Increasing			

Vertical Tangents/Cusps: a downward vertical cusp exists at $x = 0$, we can see this by obvervation:

$$\lim_{x \to x_0^+} f'(x) = \lim_{x \to 0^+} \frac{(x-2)^2(x+4)}{x^3} = +\infty$$

and

$$\lim_{x \to x_0^-} f'(x) = \lim_{x \to 0^-} \frac{(x-2)^2(x+4)}{x^3} = -\infty$$

Now to graph the function:

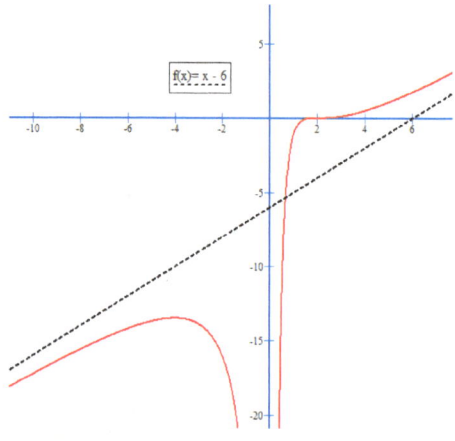

Absolute Extrema | **3.4**

Similar to finding the relative max and min, the absolute extrema are also important. Below are a few steps outlined to finding, if any, the absolute extrema of a function:

Methodology for finding Absolute Extrema:

- If a function f is continuous on a closed interval, $[a, b]$, then f has an absolute max and min, the steps for finding the absolute extrema are below:
  1. Find the critical points in the interval
  2. Determine the value of the function at all critical points within the interval
  3. Compare these values to the values of the function at a and b, the largest and smallest of the values correspond to the absolute extrema
- If a function is on an open interval, you must determine both the value of the function at its critical points as well as the end behaviour of the function. Due diligence will then be necessary in determining the absolute extrema of the function.

Example 1

Question: Find the absolute extrema of the function

$$a)\ f(x) = x^3 - 5x;\ [-2, 3],\quad b)\ g(x) = x^4 + x;\ [-2, 1]$$

Solution:

a)
Critical Points:

$$f'(x) = 3x^2 - 5 \ \rightarrow \ x = \pm\sqrt{5/3}$$

$$f(-2) = 2 \qquad f\left(\sqrt{5/3}\right) = -4.3033$$

$$f(3) = 12 \qquad f\left(-\sqrt{5/3}\right) = 4.3033$$

$\therefore$ Absolute Min = -4.3, Absolute Max = 12.

b)
Critical Points:
$$g'(x) = 4x^3 + 1 \rightarrow x = -\frac{1}{2^{(2/3)}}$$

$$g(-2) = 14 \qquad g(1) = 2$$

$$g\left(-\frac{1}{2^{(2/3)}}\right) = -0.4725$$

∴ Absolute Min = -0 .47, Absolute Max = 14.

Tips & Common Mistakes

- The reason why we check for absolute maxima and minima at the end points of the interval is because there is a very high likely hood that this could be one or both of the possible absolute extrema.

Optimization | **3.5**

This section discusses how the methods for finding absolute extrema can be applied to solve various optimization problems. Optimization problems often involve maximizing or minimizing a certain property. The following steps below can be used to tackle optimization problems:

Framework for Solving Optimization Problems:

1. Draw and label an appropriate figure including quantities and variables relevant to the problem
2. Obtain a formula to be optimized (ie. maximized or minimized), using only variables
3. Express the quantity to be optimized as a function of one variable, eliminate variables from step two by substituting known values
4. Impose a practical interval for the function based on real life limitations
5. Apply the techniques from Chapter 3.4 to obtain the maximum or minimum value of the function

Example 1

Question: Find the dimensions of a rectangle with an area of 100cm^2 whose perimeter is a minimum.

Solution:

Based on the following figure, we can construct a formula for the perimeter as follows:

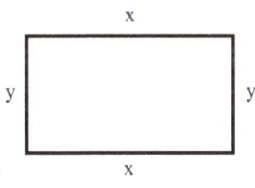

$$P = 2x + 2y$$

Now to reduce the formula down to one variable we, can use the fact that we know the area is 100cm^2:

$$100 = xy \ \rightarrow \ y = \frac{100}{x}$$

Now substituting this known constraint in our formula for the perimeter we can reduce the formula down to one variable:

$$P = 2x + \frac{200}{x}$$

Now to minimize the perimeter, we take the derivative of the function and find its critical points.

$$P' = 2 - \frac{200}{x^2}$$
$$0 = 2 - \frac{200}{x^2}$$
$$x = \pm 10$$

Therefore because we cannot have a negative side length, the minimum x value is 10cm. If we sub this value back into our function for y, $y = \frac{100}{x}$, we see that y as well has a side length of 10cm. Thus our minimized perimeter is 40cm in total.

Example 2

Question: A closed cylindrical can is to hold 800cm³ of liquid. Determine the dimensions (height and radius) to minimize the amount of material needed.

Solution:

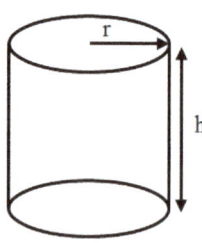

So for this question we are trying to minimize the surface area of the can, that is, the area of the top and bottom of the can ($2\pi r^2$) as well as the walls of the can ($2\pi rh$), thus we can set up the following function to minimize:

$$S = 2\pi r^2 + 2\pi rh$$

We can reduce this function down to one variable by incorporating the volume of the cylinder, which we know to be 800cm³:

$$800 = h\pi r^2$$
$$h = \frac{800}{\pi r^2}$$

Our surface area formula thus simplifies too:

$$S = 2\pi r^2 + 2\pi r \left(\frac{800}{\pi r^2}\right)$$
$$S = 2\pi r^2 + \frac{1600}{r}$$

Now we can derive this function and find its critical points to minimize its surface area:

$$S' = 4\pi r - \frac{(1600)}{r^2}$$
$$0 = 4\pi r - \frac{(1600)}{r^2}$$

$$0 = 4\pi r^3 - 1600$$
$$1600 = 4\pi r^3$$
$$r^3 = \frac{(1600)}{4\pi}$$
$$r = \sqrt[3]{\frac{1600}{4\pi}}$$
$$r \approx 5.0308cm$$

Now we can solve for the height of the can as well:

$$h \approx \frac{800}{\pi(5.0308)^2}$$
$$h \approx 10.06cm$$
$$h = 2r$$

Therefore we have found the dimensions of the can that minimize the amount of surface area, the dimensions are a radius of 5cm and a height of 10cm.

The Indefinite Integral | **4.1**

An integral is essentially the exact opposite of a derivative; and it is for this reason that integration is also known as ***antidifferentiation***. Whereas to differentiate a function is to find its rate of change over a certain period; the end goal of integration is to find a function's ***"area under the curve"***. The "area under the curve" can ultimately be related back to practical applications such as finding length, volume, density, work etc.

If a function is defined as the letter "f"; then its integral, or antiderivative, is the capital of that letter, in this case "F".

> **Definition:**
>
> $$F'(x) = f(x) \quad OR \quad \frac{d}{dx}F(x) = f(x)$$
>
> **Integral Notation:**
>
> $$\int f(x)\,dx = F(x) + C$$

As the name antiderivative suggests, at the most basic level to integrate a function is to do the exact opposite of when finding the derivative:

$$f(x) = x^4 + 1$$
$$f'(x) = 4x^{4-1}$$
$$f'(x) = 4x^3$$

$$f(x) = 4x^3$$
$$F(x) = {}^4\!/_{(3+1)}\, x^{3+1} + C$$
$$F(x) = x^4 + C$$

You will start to notice that a constant, denoted by the letter "C," keeps coming up when doing integration. You can see this in the example above. This is because when you take the derivative of a function, any constant will be eliminated (a constant will shift the function up and down the Y-axis). Thus when you perform integration there is no way of knowing what the initial constant is, so we represent it with a variable. This leads to an interesting concept known as ***integral curves*** which will be covered later in this section.

Integration Formulas

Provided below are some of the most basic and common integration formulas. Notice these specific formulas can be obtained by directly differentiating the end result.

42

$$\int dx = x + C$$

$$\int x^r dx = \frac{1}{r+1}x^{r+1} + C, (r \neq -1)$$

$$\int \cos x \, dx = \sin x + C$$

$$\int \sin x \, dx = -\cos x + C$$

$$\int e^x dx = e^x + C$$

$$\int a^x dx = \frac{a^x}{\ln a} + C \ (0 < a, a \neq 1)$$

Integral Properties

Some basic properties of integrals are summarized below:

$$\int c\, f(x)\, dx = c \int f(x)\, dx$$
$$\int [f(x) + g(x)]\, dx = \int f(x)\, dx + \int g(x)\, dx$$
$$\int [f(x) - g(x)]\, dx = \int f(x)\, dx - \int g(x)\, dx$$

Integral Curves and Slope Fields

Integral curves and Slope Fields, as seen below, occur as a result of the "+ C" that must be tagged onto each integral. This causes a vertical translation to occur to all integrals as the constant is unknown. These Integral Curves and Slope Fields try to help you visualize the infinite number of possible functions by varying the constant value "C" a finite number of times.

For example, the integral curve for $\frac{dy}{dx} = x^3$ is denoted by $y = \frac{1}{4}x^4 + C$, such that $y = \frac{1}{4}x^4$ is one integral curve and $y = \frac{1}{4}x^4 + 1$ is another. The integral curve is pictured on the left. The slope field is pictured on the right. See below.

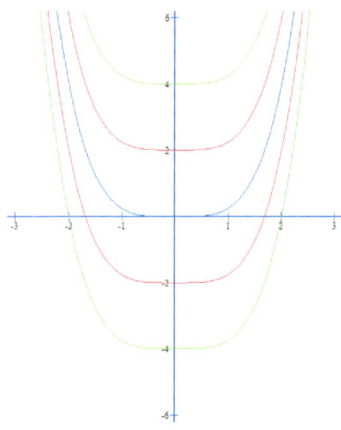

Integral curves of $y = \frac{1}{4}x^4 + C$

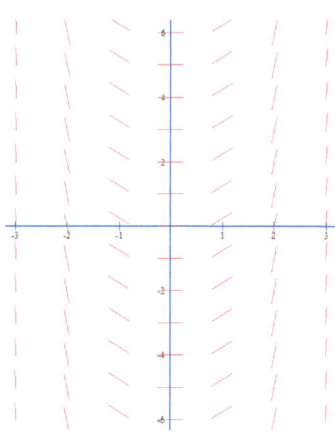

Slope field of $\frac{dy}{dx} = x^3$

Examples

Evaluate each integral:

a) $\displaystyle\int 4x + \frac{3}{2x^4}\, dx,$ b) $\displaystyle\int 4e^{2x} + 8^x\, dx,$ c) $\displaystyle\int \sqrt{x}(4-x)^2\, dx$

Solution a):

$$= \int 4x + \frac{3}{2x^4}\, dx$$

$$= \int 4x\, dx + \int \frac{3}{2x^4}\, dx$$

$$= \frac{4}{1+1}x^{1+1} + C + \int \frac{3}{2}x^{-4}\, dx$$

$$= 2x^2 + \frac{3}{2}\frac{1}{-4+1}x^{-4+1} + C$$

$$= 2x^2 + \frac{3}{2}\frac{1}{-3}x^{-3} + C$$

$$= 2x^2 + \frac{3}{-6}x^{-3} + C$$

$$= 2x^2 - \frac{3}{6}x^{-3} + C$$

$$\vdots$$

Solution b):

$$= \int 4e^{2x} + 8^x\, dx$$

$$= \int 4e^{2x}\, dx + \int 8^x\, dx$$

$$= \frac{4}{2}e^{2x} + \frac{8^x}{\ln 8} + C$$

$$= 2e^{2x} + \frac{8^x}{\ln 8} + C$$

$$= 2x^2 - \frac{3}{6x^3} + C$$
$$= 2x^2 - \frac{1}{2x^3} + C$$

Solution *c)*:

$$= \int \sqrt{x}(4-x)^2 \, dx$$
$$= \int \sqrt{x}(4-x)(4-x) \, dx$$
$$= \int \sqrt{x}(x^2 - 8x + 16) \, dx$$
$$= \int x^{2.5} - 8x^{1.5} + 16x^{0.5} \, dx$$
$$= \int x^{\frac{5}{2}} - 8x^{\frac{3}{2}} + 16x^{\frac{1}{2}} \, dx$$
$$= \frac{1}{\frac{5}{2}+1} x^{\frac{5}{2}+1} - \frac{8}{\frac{3}{2}+1} x^{\frac{3}{2}+1} + \frac{16}{\frac{1}{2}+1} x^{\frac{1}{2}+1} + C$$
$$= \frac{2}{5+2} x^{\frac{7}{2}} - \frac{8 \cdot 2}{3+2} x^{\frac{5}{2}} + \frac{16 \cdot 2}{1+2} x^{\frac{3}{2}} + C$$
$$= \frac{2}{7} x^{\frac{7}{2}} - \frac{16}{5} x^{\frac{5}{2}} + 16x^{\frac{3}{2}} + C$$
$$= \frac{2}{7} \sqrt{x^7} - \frac{16}{5} \sqrt{x^5} + \frac{32}{3} \sqrt{x^3} + C$$

Tips & Common Mistakes

- Remember that the function used to generate the Slope Field is always the derivative of the Integral Curve. Therefore, the integral curve is always the integral of the function used to generate the slope field.

- Do not get mixed up with the following:

$$Correct: \int x^r \, dx = \frac{x^{r+1}}{r+1} + C = \frac{1}{r+1} x^{r+1} + C$$

$$Incorrect: \int x^r \, dx = \left(\frac{x}{r+1}\right)^{r+1} + C$$

U-Substitution | 4.2

Substitution, often dubbed *U-substitution*, is a method of integration which simplifies a complex function so that it can be more easily recognized and integrated. Follow the methodology below:

If an integral is of the form:

$$\int f[g(x)]g'(x)\,dx = F[g(x)] + C$$

Such that: $u = g(x)$
               $du = g'(x)\,dx$

Then substitute:

$$\int f[g(x)]g'(x)\,dx = F[g(x)] + C$$
$$\int f[\ u\ ]\quad du\quad = F[\ u\ ] + C$$

Example 1

Question: Evaluate $\int 2x(x^2 + 4)^{25}\,dx$

Solution:

$u = x^2 + 4$
$du = 2x \cdot dx$

And now substituting this into the integral:

$$= \int (u)^{25}\,du, \quad \text{which is now a much easier integral}$$

$$= \frac{u^{26}}{26} + C, \quad \text{now substitute back in the value of } u$$

$$= \int 2x(x^2 + 4)^{25}$$

$$= \frac{(x^2 + 4)^{26}}{26} + C$$

$$\therefore \int 2x(x^2 + 4)^{25} dx = \frac{(x^2 + 4)^{26}}{26} + C$$

Example 2

Question: Evaluate

$$a) \int \sin(x + 8)\, dx, \qquad b) \int (\cos^4 \theta) \sin \theta\, d\theta, \qquad c) \int \sec^2 \pi x\, d\theta$$

Solution:

a)

$u = x + 8$

$du = 1 \cdot dx$

$= \int \sin(u)\, du$

$= -\cos u + C$

$= -\cos(x + 8) + C$

b)

rewrite as: $\int (\cos \theta)^4 \sin \theta\, d$

$u = \cos \theta$

$du = -\sin\theta \cdot d\theta$

$-du = \sin\theta \cdot d\theta$

$= -\int u^4 du$

$= -\frac{u^5}{5} + C$

$= -\frac{\cos^5 \theta}{5} + C$

c)

rewrite as: $\int (\sec \pi\theta)^2 d\theta$

$u = \pi\theta$

$du = \pi \cdot d\theta$

$\frac{1}{\pi} \cdot du = d\theta$

$= \frac{1}{\pi} \int (\sec u)^2 du$

$= \frac{1}{\pi} \tan u + C$

$= \frac{1}{\pi} \tan \pi\theta + C$

Example 3

Question: Evaluate:

$$a) \int x^2 \sqrt{x - 1}\, dx, \qquad b) \int \cos^3 \theta\, d\theta$$

Solution:

a)

$u = x - 1$

$du = 1 \cdot dx$

$= \int x^2 \sqrt{u}\, du$, but there is still an x^2 term

Using the equality: $u = x - 1$

$x = u + 1$

$x^2 = (u + 1)^2$

$= \int (u + 1)^2 \sqrt{u}\, du$

$= \int (u^2 + 2u + 1)\sqrt{u}\, du$

$= \int \left(u^{5/2} + 2u^{3/2} + u^{1/2}\right) du$

$= \frac{2}{7}u^{7/2} + \frac{4}{5}u^{5/2} + \frac{2}{3}u^{3/2} + C$

$= \frac{2}{7}(x - 1)^{7/2} + \frac{4}{5}(x - 1)^{5/2} + \frac{2}{3}(x - 1)^{3/2} + C$

Solution:

b)

both $u = \cos\theta$ or $u = \cos^2\theta$

will not work, you can verify.

$\therefore$ *rewrite as:* $\int \cos^2\theta \cos\theta\, d\theta$

$= \int \cos^2\theta \cos\theta\, d\theta$

$= \int (1 - \sin^2\theta) \cos\theta\, d\theta$

$u = \sin\theta$

$du = \cos\theta \cdot d\theta$

$= \int (1 - u^2)du$

$= u - \frac{u^3}{3} + C$

$= u - \frac{u^3}{3} + C$

$= \sin\theta - \frac{\sin^3\theta}{3} + C$

Tips & Common Mistakes

- One trick to recognize if the integration problem can be solved by substitution is to look for things that are derivatives of each other, that way the function can be easily simplified

Integration by Parts | 4.3

Another method of integral evaluation is the integration by parts method. This method applies when the integral is a product of two different functions, it can then be represented as seen below:

$$\int f(x)g(x)\,dx = f(x)G(x) - \int f'(x)G(x)\,dx$$

Although it is more commonly written as seen below, the values of u and dv must be chosen:

$$\int u\,dv = uv - \int v\,du$$

$$\downarrow\ \begin{array}{l} u = f(x) \\ du = f'(x)dx \end{array} \qquad \uparrow\ \begin{array}{l} v = G(x) \\ dv = g(x)dx \end{array}$$

For a Definite Integral (covered in Chapter 4.4):

$$\int_a^b u\,dv = uv]_a^b - \int_a^b v\,du$$

Example 1

Question: Evaluate:

$$a)\ \int x \cos x\ dx$$

Solution:

Chose first $u = x$ and $dv = \cos x\ dx$, we then derive u and integrate dv:

$$\downarrow\ \begin{array}{l} u = x \\ du = ? \end{array} \qquad \uparrow\ \begin{array}{l} v = ? \\ dv = \cos x\,dx \end{array}$$

$$\downarrow\ \begin{array}{l} u = x \\ du = 1 \cdot dx \end{array} \qquad \uparrow\ \begin{array}{l} v = \sin x \\ dv = \cos x\,dx \end{array}$$

Therefore:

$$\int x \cos x \ dx = x \sin x - \int \sin x \ dx$$

$$= x \sin x - (-\cos x) + C$$

$$= x \sin x + \cos x + C$$

Example 2

Question: Evaluate:

$$a) \int xe^{-x} \ dx, \qquad b) \int \ln x \ dx$$

Solution:

a)
Choose $u = x$ and $dv = e^{-x} \ dx$

$$\downarrow \begin{array}{l} u = x \\ du = 1 \cdot dx \end{array} \qquad \uparrow \begin{array}{l} v = -e^{-x} \\ dv = e^{-x} \ dx \end{array}$$

$$= -xe^{-x} - \int (-e^{-x}) \ dx$$

$$= -xe^{-x} + \int e^{-x} \ dx$$

$$= -xe^{-x} - e^{-x} + C$$

b)
Rewrite as: $\int \ln x \cdot 1 dx$

Choose $u = \ln x$ and $dv = 1 \cdot dx$

$$\downarrow \begin{array}{l} u = \ln x \\ du = \frac{1}{x} dx \end{array} \qquad \uparrow \begin{array}{l} v = x \\ dv = 1 \cdot dx \end{array}$$

$$= x \ln x - \int x \frac{1}{x} dx$$

$$= x \ln x - \int 1 \cdot dx$$

$$= x \ln x - x + C$$

Example 3

Question: Evaluate $a) \int e^{2x} \sin x \ dx$

Solution:
Chose first $u = e^{2x}$ and $dv = \sin x \ dx$:

$$\downarrow \begin{array}{l} u = e^{2x} \\ du = 2e^{2x} dx \end{array} \uparrow \begin{array}{l} v = -\cos x \\ dv = \sin x \ dx \end{array}$$

$$= -e^{2x} \cos x + 2 \int e^{2x} \cos x \ dx$$

** We must now carry out a second integration by parts on

$$\int e^{2x} \cos x \, dx$$

and sub it back into the integral, as seen to the right:

$$\int e^{2x} \cos x \, dx$$

Chose first $u = e^{2x}$ and $dv = \cos x \ dx$:

$$\downarrow u = e^{2x} \qquad \uparrow v = \sin x$$
$$du = 2e^{2x} dx \qquad dv = \cos x \, dx$$

$$= e^{2x} \sin x - 2 \int e^{2x} \sin x \, dx$$

$$= -e^{2x} \cos x + 2\left[e^{2x} \sin x - 2 \int e^{2x} \sin x \, dx \right]$$
$$= -e^{2x} \cos x + 2e^{2x} \sin x - 4 \int e^{2x} \sin x \, dx$$

Seeing as how after our second integration by parts we still have the original integral in our equation, you may be thinking that the integration by parts method is not appropriate for this integral. But we can still relate this final term of $-e^{2x} \cos x + 2e^{2x} \sin x - 4 \int e^{2x} \sin x \, dx$ back to our original integral, as seen below:

$$\int e^{2x} \sin x \ dx = -e^{2x} \cos x + 2e^{2x} \sin x - 4 \int e^{2x} \sin x \, dx$$

We can group the two integrals together!

$$5 \int e^{2x} \sin x \ dx = -e^{2x} \cos x + 2e^{2x} \sin x$$
$$\int e^{2x} \sin x \ dx = \frac{1}{5}(-e^{2x} \cos x + 2e^{2x} \sin x) + C$$

Tips & Common Mistakes

- Here is a useful strategy for decided which function is u and which is dv, it is referred to as the LIATE method:

  Logarithmic, **I**nverse Trigonometric, **A**lgebraic, **T**rigonometric, **E**xponential

  You will often be successful if you take u to be the function that occurs first in the list.

The Fundamental Theorem of Calculus | 4.4

The Definite Integral

The Fundamental Theorem of Calculus is used to calculate the area under the curve between two intervals. It can be visually described by using rectangles of an ever decreasing base. As the base of these rectangles, denoted by Δx, approaches zero, the area approximation will approach the true value across the given interval. Follow the example below as it gets ever closer to approximating the true value of $y = \sin x$ across the interval of 0 to π:

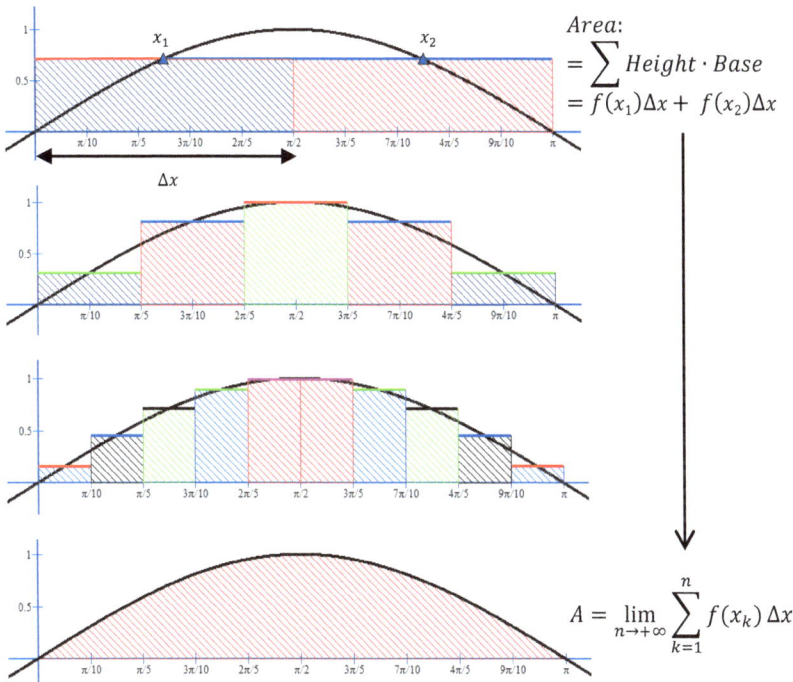

$$Area:$$
$$= \sum Height \cdot Base$$
$$= f(x_1)\Delta x + f(x_2)\Delta x$$

$$A = \lim_{n \to +\infty} \sum_{k=1}^{n} f(x_k)\,\Delta x$$

This is the base of the Fundamental Theorem of Calculus, from this it was derived the final theorem which we will cover in this chapter, and which you will use to calculate the exact area integral over two intervals. Calculating area over two intervals is known as the ***Definite Integral***.

53

The Fundamental Theorem of Calculus

The area can be evaluated by finding the antiderivative at the upper interval (usually denoted by the variable b), and then subtracting that from the antiderivative at the lower interval (usually denoted by the variable a).

If f is continuous on [a, b] and F is any antiderivative of f on [a, b], then

$$\int_a^b f(x)dx = F(x)\Big|_a^b = F(b) - F(a)$$

The upper limit can also be variable, for which we denote the upper limit as "x:"

$$\int_a^x f(t)dt = F(x) - F(a)$$

One interesting thing to note with the Definite Integral is that the **constants subtract out of the final equation**. See the arbitrary example below:

$$\int_a^b f(x)dx = F(x)\Big|_a^b = [F(b) + C] - [F(a) + C]$$

$$= F(b) - F(a) + C - C$$

$$= F(b) - F(a)$$

Example 1

Evaluate:

$$\int_1^2 x^3 \, dx$$

Solution:

$$= \int_1^2 x^3 \, dx$$

$$= \frac{1}{4}x^4\Big|_1^2$$

$$= \frac{1}{4}(2)^4 - \frac{1}{4}(1)^4$$

$$= \frac{1}{4}16 - \frac{1}{4}1$$

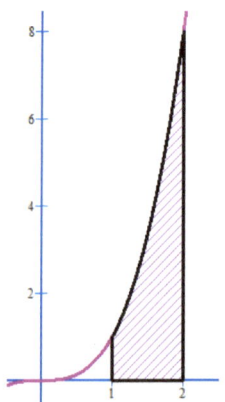

$$= 4 - \frac{1}{4}$$
$$= 3.75$$

Example 2

Evaluate:

$$\int_0^\pi \cos x \ dx$$

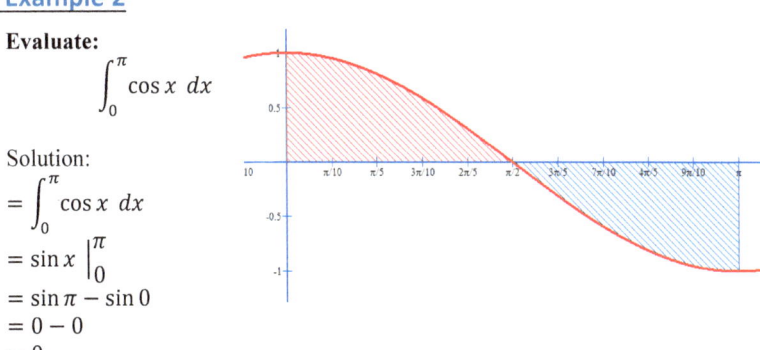

Solution:

$$= \int_0^\pi \cos x \ dx$$
$$= \sin x \ \Big|_0^\pi$$
$$= \sin \pi - \sin 0$$
$$= 0 - 0$$
$$= 0$$

You might be pondering why this solution yields an answer of zero. This example was specifically designed to show you that areas below the x-axis are negative areas. Therefore because the red shaded area and the black shaded area are of equal value, they will subtract each other out. ***Watch out for this!***

U-Substitution and the Definite Integral

One interesting thing to note when doing U-Substitution with a Definite Integral is that the **limits of integration will change with the substitution**. Do not fret though, it is a simple procedure!

All you must do is plug the old limits into your new U – equation, these will give you your new limits. Please see the example below for the following definite integral:

$$= \int_1^2 x^2 (x^3 + 1)^4 \ dx$$

Using the equality: ***New Limits*** ⟶
$$u = x^3 + 1$$
$$du = 3x^2 dx$$
$$\frac{1}{3} du = x^2 dx$$

New Lower Limit:
$$u = x^3 + 1$$
$$= (Old \ lower \ Limit)^3 + 1$$
$$= (1)^3 + 1$$
$$= 2$$
New Upper Limit:
$$u = x^3 + 1$$
$$= (Old \ Upper \ Limit)^3 + 1$$
$$= (2)^3 + 1$$
$$= 9$$

Now we substitute our new limits, u and $\frac{1}{3} du$ in:

$$= \frac{1}{3} \int_{2}^{9} (u)^4 \; du$$

$$= \frac{1}{3} \left[\frac{1}{4+1} (u)^5 \right]_{2}^{9}$$

$$= \frac{1}{3} \left[\frac{1}{5} (u)^5 \right]_{2}^{9}$$

$$= \frac{1}{3} \left[\frac{1}{5} (9)^5 - \frac{1}{5} (2)^5 \right]$$

$$= \frac{9^5 - 2^5}{15}$$

Closing Thoughts

Practice makes perfect, the more questions and past tests you do the higher your final mark will be. You've probably heard this a million times by now, but more practice really does correlate to a greater understanding of concepts and thus higher marks.

I wish you good luck in your course and studying for quizzes, tests and exams.

***

I love to hear about the success of my readers. Tell me about how you were able to understand tough concepts and other successes. Please send an email to feedback.mathbridge@gmail.com.

Notes

Notes

Notes

Notes

Notes

Notes

www.ingramcontent.com/pod-product-compliance
Lightning Source LLC
Chambersburg PA
CBHW040834180526
45159CB00001B/184